THE CURSED WEPON

The history of biological warfare

Navneet Kumar

Copyright © 2021 Navneet kumar

All rights reserved

The characters and events portrayed in this book are fictitious. Any similarity to real persons, living or dead, is coincidental and not intended by the author.

No part of this book may be reproduced, or stored in a retrieval system, or transmitted in any form or by any means, electronic, mechanical, photocopying, recording, or otherwise, without express written permission of the publisher.

ISBN-13: 9798515073763
ISBN-10: 1477123456

Cover design by: Art Painter
Library of Congress Control Number: 2018675309
Printed in the United States of America

THE HISTORY OF BIOLOGICAL WARFARE

Human experimentation, modern nightmares and lone madmen in the twentieth century

During the past century, more than 500 million people died of infectious diseases. Several tens of thousands of these deaths were due to the deliberate release of pathogens or toxins, mostly by the Japanese during their attacks on China during the Second World War. Two international treaties outlawed biological weapons in 1925 and 1972, but they have largely failed to stop countries from conducting offensive weapons research and large-scale production of biological weapons. And as our knowledge of the biology of disease-causing agents—viruses, bacteria, and toxins— increases, it is legitimate to fear that modified pathogens could constitute devastating agents for biological warfare. To put these future threats into perspective, I discuss in this article the history of biological warfare and terrorism.

> During the [Second World War], the Japanese army poisoned more than 1,000 water wells in Chinese villages to study cholera and ty-

phus outbreaks

Man has used poisons for assassination purposes ever since the dawn of civilization, not only against individual enemies but also occasionally against armies (Table 1). However, the foundation of microbiology by Louis Pasteur and Robert Koch offered new prospects for those interested in biological weapons because it allowed agents to be chosen and designed on a rational basis. These dangers were soon recognized, and resulted in two international declarations—in 1874 in Brussels and 1899 in The Hague—that prohibited the use of poisoned weapons. However, although these, as well as later treaties, were all made in good faith, they contained no means of control, and so failed to prevent interested parties from developing and using biological weapons. The German army was the first to use weapons of mass destruction, both biological and chemical, during the First World War, although their attacks with biological weapons were on a rather small scale and were not particularly successful: covert operations using both anthrax and glanders (Table 2) attempted to infect animals directly or to contaminate animal feed in several of their enemy countries (Wheelis, 1999). After the war, with no lasting peace established, as well as false and alarming intelligence reports, various European countries instigated their biological warfare programs, long before the onset of the Second World War (Geissler & Moon, 1999).

Table 1 | Examples of biological warfare during the past millennium

Year	Event
1155	Emperor Barbarossa poisons water wells with human bodies, Tortona, Italy
1346	Mongols catapult bodies of plague victims over the city walls of Caffa, Crimean Peninsula
1495	Spanish mix wine with blood of leprosy patients to sell to their French foes, Naples, Italy
1650	Polish fire saliva from rabid dogs towards their enemies
1675	First deal between German and French forces not to use 'poison bullets'
1763	British distribute blankets from smallpox patients to native Americans
1797	Napoleon floods the plains around Mantua, Italy, to enhance the spread of malaria
1863	Confederates sell clothing from yellow fever and smallpox patients to Union troops, USA

In North America, it was not the government but a dedicated individual who initiated a bioweapons research program. Sir Frederick Banting, the NobelPrize-winning discoverer of insulin, created what could be called the first private biological weapon research center in 1940, with the help of corporate sponsors (Avery, 1999; Regis, 1999). Soon afterward, the US government was also pressed to perform such research by their British allies who, along with the French, feared a German attack with biological weapons (Moon, 1999, Regis, 1999), even though the Nazis never seriously considered using biological weapons (Geissler, 1999). However, the Japanese embarked on a large-scale program to develop biological weapons during the Second World War (Harris, 1992, 1999, 2002) and eventually used them in their conquest of China. Indeed, alarm bells should have rung as early as 1939, when the Japanese legally, and then illegally, attempted to obtain the yellow fever virus from the Rockefeller Institute in New York (Harris, 2002).

Table 2 | Crucial biological agents (Centers for Disease Control, Atlanta, Georgia, USA)

Disease	Pathogen	Abused[1]
Category A (major public health hazards)		
Anthrax	*Bacillus antracis* (B)	First World War
		Second World War
		Soviet Union, 1979
		Japan, 1995
		USA, 2001
Botulism	*Clostridium botulinum* (T)	–
Haemorrhagic fever	Marburg virus (V)	Soviet bioweapons programme
	Ebola virus (V)	–
	Arenaviruses (V)	–
Plague	*Yersinia pestis* (B)	Fourteenth-century Europe
		Second World War
Smallpox	*Variola major* (V)	Eighteenth-century N. America
Tularemia	*Francisella tularensis* (B)	Second World War
Category B (public health hazards)		
Brucellosis	*Brucella* (B)	–
Cholera	*Vibrio cholerae* (B)	Second World War
Encephalitis	Alphaviruses (V)	Second World War
Food poisoning	*Salmonella, Shigella* (B)	Second World War
		USA, 1990s
Glanders	*Burkholderia mallei* (B)	First World War
		Second World War
Psittacosis	*Chlamydia psittaci* (B)	–
Q fever	*Coxiella burnetti* (B)	–
Typhus	*Rickettsia prowazekii* (B)	Second World War
Various toxic syndromes	Various bacteria	Second World War

The father of the Japanese biological weapons program, the radical nationalist Shiro Ishii, thought that such weapons would constitute formidable tools to further Japan's imperialistic plans. He started his research in 1930 at the Tokyo Army Medical School and later became head of Japan's bioweapon program during the Second World War (Harris, 1992, 1999, 2002). At its height, the program employed more than 5,000 people and killed as many as 600 prisoners a year in human experiments in just one of its 26 centers. The Japanese tested at least 25 different disease-causing agents on prisoners and unsuspecting civilians. During the war, the Japanese army poisoned more than 1,000 water wells in Chinese villages to study cholera and typhus outbreaks. Japanese planes dropped plague-infested fleas over Chinese cities or distributed them employing saboteurs in rice fields and along roads. Some of the epidemics they caused persisted for years and continued to kill more than 30,000 people in 1947, long after the Japanese had surrendered (Harris, 1992, 2002). Ishii's troops also used some of their agents against the Soviet army, but it is unclear as to whether the casualties on both sides were caused by this deliberate spread of disease or by natural infections (Harris, 1999). After the war, the Soviets convicted some of the Japanese biowarfare researchers for war crimes, but the USA granted freedom to all researchers in exchange for information on their human experiments. In this way, war criminals once more became respected citizens, and some went on to found pharmaceutical companies. Ishii's successor, Masaji Kitano, even published postwar research articles on human experiments, replacing 'human' with 'monkey' when referring to the experiments in wartime China (Harris, 1992, 2002).

Although some US scientists thought the Japanese information insightful, it is now largely assumed that it was of no real help to the US biological warfare program projects. These started in 1941 on a small scale but increased during the war to include more than 5,000 people by 1945. The main effort focused on developing capabilities to counter a Japanese attack with biological weapons, but documents indicate that the US government also discussed the offensive use of anticrop weapons (Bernstein, 1987). Soon after the war, the US military started open-air tests, exposing test animals, human volunteers, and unsuspecting civilians to both pathogenic and non-pathogenic microbes (Cole, 1988; Regis, 1999). A release of bacteria from naval vessels off the coasts of Virginia

> ....nobody really knows what the Russians are working on today and what happened to the weapons they produced

and San Francisco infected many people, including about 800,000 people in the Bay area alone. Bacterial aerosols were released at more than 200 sites, including bus stations and airports. The most infamous test was the 1966 contamination of the New York metro system with Bacillus globigii—a non-infectious bacterium used to simulate the release of anthrax—to study the spread of the pathogen in a big city. But with the opposition to the Vietnam War growing and the realization that biological weapons could soon become the poor man's nuclear bomb, President Nixon decided to abandon offensive biological weapons research and signed the Biological and Toxin Weapons Convention (BTWC) in 1972, an improvement on the 1925 Geneva Protocol. Although the latter disallowed only the use of chemical or biological weapons, the BTWC also prohibits research on biological weapons. However, the BTWC does not include means for verification, and it is somewhat ironic that the US administration let the verification protocol fail in 2002, particularly because of the Soviet bioweapons project, which not only was a clear breach of the BTWC, but also remained undetected for years.

Even though they had just signed the BTWC, the Soviet Union established Biopreparat, a gigantic biowarfare project that, at its height, employed more than 50,000 people in various research and production centers (Alibek & Handelman, 1999). The size and scope of the Soviet Union's efforts were truly staggering: they produced and stockpiled tons of anthrax bacilli and smallpox virus, some for use in intercontinental ballistic missiles, and engineered multidrug-resistant bacteria, including plague. They worked on hemorrhagic fever viruses, some of the deadliest pathogens that humankind has encountered. When virologist Nikolai Ustinov died after injecting himself with the deadly Marburg virus, his colleagues, with the mad logic and enthusiasm of bioweapon developers, re-isolated the virus from his body and found that it had mutated into a more virulent form than the one that Ustinov had used. And few took any notice, even when accidents happened. In 1971, smallpox broke out in the Kazakh city of Aralsk and killed three of the ten people that were infected. It is speculated that they were infected from a bioweapons research center on a small island in the Aral Sea (Enserink, 2002). In the same area, on other occasions, several fishermen and a researcher died from plague and glanders, respectively (Miller et al., 2002). In 1979, the Soviet secret police orchestrated a large cover-up to explain an outbreak of anthrax in Sverdlovsk, now Ekaterinburg, Russia, with poisoned meat from anthrax-contaminated animals sold on the black market. It was eventually revealed to have been due to an accident in a bioweapons factory, where a clogged air filter was removed but not replaced between shifts (Fig. 1) (Meselson et al., 1994; Alibek & Handelman, 1999).

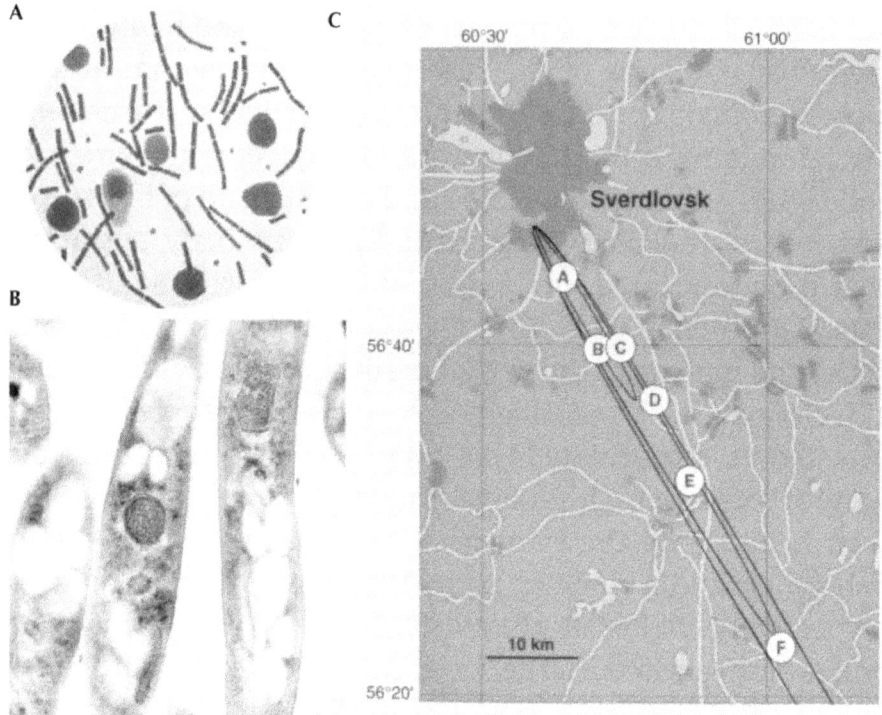

Fig. 1 | Anthrax as a biological weapon. Light (A) and electron (B) micrographs of anthrax bacilli, reproduced from the Centers of Disease Control Public Health Image Library. The map (C) shows six villages in which animals died after anthrax spores were released from a bioweapons factory in Sverdlovsk, USSR, in 1979. Settled areas are shown in grey, roads in white, lakes in blue and the calculated contours of constant dosage of anthrax spores in black. At least 66 people died after the accident. (Reprinted with permission from Meselson et al., 1994 Â© (1994) American Association for the Advancement of Science.)

The most striking feature of the Soviet program was that it remained secret for such a long time. During the Second World War, the Soviets used a simple trick to check whether US researchers were occupied with secret research: they monitored whether American physicists were publishing their results. Indeed, they were not, and the conclusion was, correctly, that the US was busy building a nuclear bomb

(Rhodes, 1988, pp. 327 and 501). The same trick could have revealed the Soviet bioweapons program much earlier (Fig. 2). With the collapse of the Soviet Union, most of these programs were halted and the research centers abandoned or converted for civilian use. Nevertheless, nobody knows what the Russians are working on today and what happened to the weapons they produced. Western security experts now fear that some stocks of biological weapons might not have been destroyed and have instead fallen into other hands (Alibek & Handelman, 1999; Miller et al., 2002). According to US intelligence, South Africa, Israel, Iraq, and several other countries have developed or still are developing biological weapons (Zilinskas, 1997; Leitenberg, 2001).

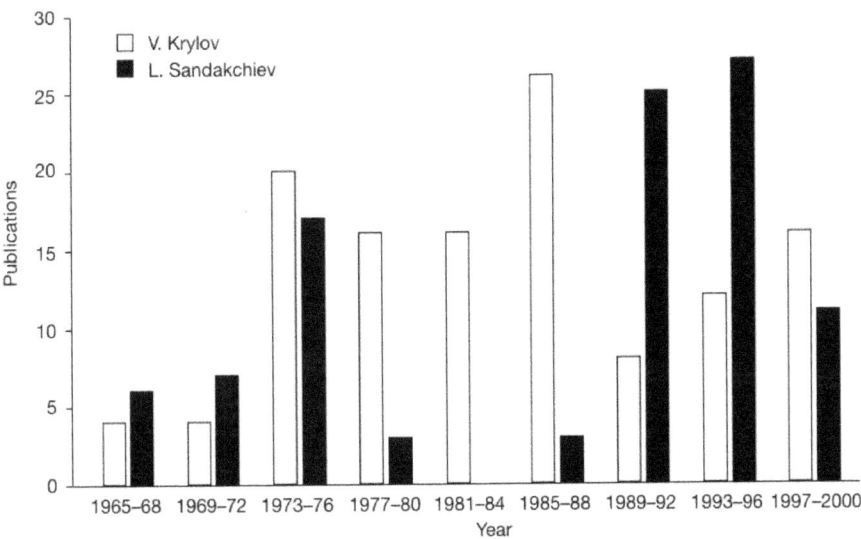

Fig. 2 | Detecting biological warfare research. A comparison of the number of publications from two Russian scientists. L. Sandakchiev (black bars) was involved, as the head of the Vector Institute for viral research, in the Soviet project to produce smallpox as an offensive biological weapon. V. Krylov (white bars) was not. Note the decrease in publications by Sandakchiev compared with those by Krylov. The data were compiled from citations from a PubMed search for the researchers on 15 August 2002.

Apart from state-sponsored biowarfare programs, individuals and non-governmental groups have also gained access to potentially dangerous microorganisms, and some have used them (Purver, 2002). A few examples include the spread of hepatitis, parasitic infections, severe diarrhea, and gastroenteritis. The latter occurred when a religious sect tried to poison a whole community by spreading Salmonella in salad bars to interfere with a local election (Török et al., 1997; Miller et al., 2002). The sect, which ran a hospital on its grounds, obtained the bacterial strain from a commercial supplier. Similarly, a right-wing laboratory technician tried to get hold of the plague bacterium from the American Tissue Culture Collection and was only discovered after he complained that the procedure took too long (Cole, 1996). These examples indicate that organized groups or individuals with sufficient determination can obtain dangerous biological agents. All that is required is a request to 'colleagues' at scientific institutions, who share their published materials with the rest of the community (Breithaupt, 2000). The relative ease with which this can be done explains why the numerous hoaxes in the USA after the anthrax mailings had to be taken seriously, thus causing an estimated economic loss of US $100 million (Leitenberg, 2001).

> *These examples clearly indicate that organized groups or individuals with sufficient determination can obtain dangerous biological agents*

Another religious cult, in Japan, proved both the ease and the difficulties of using biological weapons. In 1995, the Aum Shinrikyo cult used Sarin gas in the Tokyo subway, killing 12 train passengers and injuring more than 5,000 (Cole, 1996). Before these attacks, the sect had also tried, on several occasions, to distribute (non-infectious) anthrax within the city with no success. It was easy for the sect members to produce the spores but much harder to disseminate them (Atlas, 2001; Leitenberg, 2001). The still-unidentified culprits of the 2001 anthrax attacks in the USA were more successful, sending contaminated letters that eventually killed five people and, potentially even more seriously, caused an upsurge in demand for antibiotics, resulting in over-use and thus contributing to drug resistance (Atlas, 2001; Leitenberg, 2001; Miller et al., 2002).

Cuba frequently accused the USA of using biological warfare

One interesting aspect of biological warfare is the accusations made by the parties involved, either as excuses for their actions or to justify their political goals. Many of these allegations, although later shown to be wrong, have been exploited either as propaganda or as a pretext for war, as recently seen in the case of Iraq. It is essential to draw the line between fiction and reality, particularly if, based on such evidence, politicians call for a 'pre-emptive war or allocate billions of dollars to research projects. Examples of such incorrect allegations include a British report before the Second World War that German secret agents were experimenting with bacteria in the Paris and London subways, using harmless species to test their dissemination through the transport system (Regis, 1999; Leitenberg, 2001). Although this claim was never substantiated, it might have had a role in promoting British research on anthrax in Porton Down and on

Gruinard Island. During the Korean War, the Chinese, North Koreans, and Soviets accused the USA of deploying biological weapons of various kinds. This is now seen as wartime propaganda, but the secret deal between the USA and Japanese bioweapons researchers did not help to diffuse these allegations (Moon, 1992). Later, the USA accused the Vietnamese of dropping fungal toxins on the US Hmong allies in Laos. However, it was found that the yellow rain associated with the reported variety of syndromes was simply bee feces (Fig. 3; Seeley et al., 1985). The problem with such allegations is that they develop a life of their own, no matter how unbelievable they are. For example, the conspiracy theory that HIV is a biological weapon is still alive in some people's minds. Depending on whom one asks, KGB or CIA scientists developed HIV to damage the USA or to destabilize Cuba, respectively. Conversely, in 1997, Cuba was the first country to officially file a complaint under Article 5 of the BTWC, accusing the USA of releasing a plant pathogen (Leitenberg, 2001). Although this was never proven, the USA did indeed look into biological agents to kill Fidel Castro and Frederik Lumumba of the Democratic Republic of Congo (Miller et al., 2002).

Fig. 3 | Hmong refugees from Laos, who collaborated with the American armed forces during the Vietnam War, accused the Soviet Union of attacking them with biological or chemical weapons. However, the alleged toxin warfare agent known as yellow rain matches perfectly the yellow spots of bee faeces on leaves in the forest of the Khao Yai National Park in Thailand. (Image reprinted with permission from Seeley et al., 1985 Â© (1985) M. Meselson, Harvard University)

We are witnessing a renewed interest in biological warfare and terrorism owing to several factors, including the discovery that Iraq has been developing biological weapons (Zilinskas, 1997), several best-selling novels describing biological attacks, and the anthrax letters after the terrorist attacks on 11 September 2001. As history tells us, virtually no nation with the ability to develop weapons of mass destruction has abstained from doing so. And the Soviet project shows that international treaties are useless unless an effective verification procedure is in place. Unfortunately, the same knowledge that is needed to develop drugs and vaccines against pathogens has the potential to be abused for the development of biological weapons (Fig. 4; Finkel, 2001). Thus, some critics have suggested that information about potentially harmful pathogens should not be made public but rather put into the hands of 'appropriate representatives' (Danchin, 2002; Wallerstein, 2002). A recent report on anti-crop agents was already self-censored before publication, and journal editors now recommend special scrutiny for sensitive papers (Mervis & Stokstad, 2002; Cozzavelli, 2003; Malakoff, 2003). Whether or not such measures are useful deterrents might be questionable, because the application of available knowledge is enough to kill. An opposing view calls for the imperative publication of information about the development of biological weapons to give scientists, politicians, and the interested public all the necessary information to determine a potential threat and devise countermeasures.

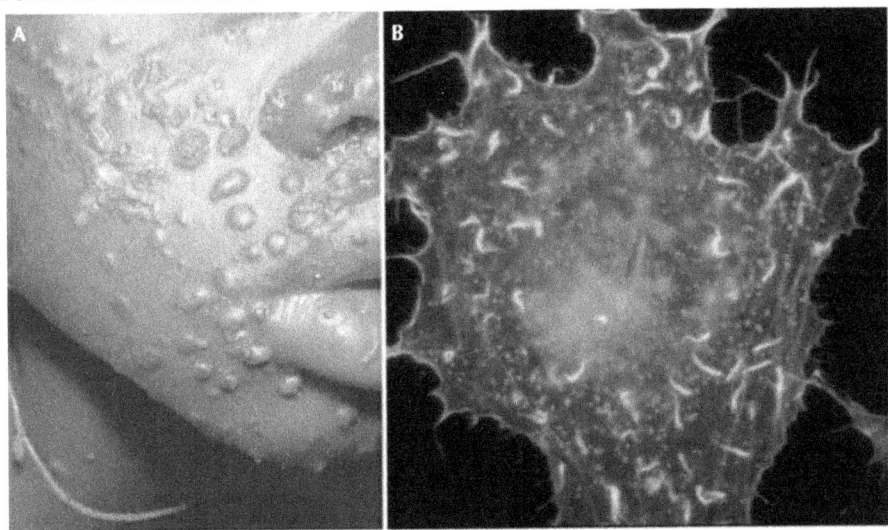

Fig. 4 | Intimate interactions of hosts and pathogens. (A) The face of a smallpox victim in Accra, Ghana, 1967. (Photograph from the Center of Disease Control's Public Health Image Library.) (B) A poxvirusinfected cell is shown to illustrate just one of the many intricate ways in which pathogens can interact with, abuse or mimic their hosts. The virus is shown in red, the actin skeleton of the cell in green. Emerging viruses rearrange actin into tail-like structures that push them into neighbouring cells. (Image by F. Frischknecht and M. Way, reprinted with permission from the Journal of General Virology.)

The current debate about biological weapons is certainly important in raising awareness and increasing our preparedness to counter a potential attack. It could also prevent an overreaction such as that caused in response to the anthrax letters mailed in the USA. However, contrasting the speculative nature of biological attacks with the grim reality of the millions of people who still die each year from preventable infections, we might ask ourselves just how many resources we can afford to allocate in preparation for a hypothetical human-inflicted disaster.

ACKNOWLEDGEMENT

I am grateful to P. Baldacci, G. Frazzetto, B. Janssens, U. Kornak and R. Menard for comments on the manuscript. My research is supported by a long-term fellowship from the Human Frontier Science Program.

REFERENCES

Alibek, K. & Handelman, S. (1999) Biohazard. Random House, New York, USA.

Atlas, R.A. (2001) Bioterrorism before and after September 11. Crit.Rev.Microbiol., 27, 355-379.

Avery, D. (1999) in Biological and Toxin Weapons: Research, Development and Use from the Middle Ages to 1945 (eds Geissler, E. & Moon, J.E.v.C.), 190-214. Stockholm International Peace Research Institute, Oxford University Press, Oxford, UK.

Bernstein, B.J. (1987) The birth of the U.S. biological-warfare program. Sci.Am., 255, 94-99.

Breithaupt, H. (2000) Toxins for terrorists. EMBO Rep., 1, 298-301.

Cole, L.A. (1988) Clouds of Secrecy:The Armys Germ Warfare Tests Over Populated Areas. Rowman & Littlefield, Lanham, Maryland, USA.

Cole, L.A. (1996) The specter of biological weapons. Sci.Am., 275, 30-35.

Cozzavelli, N. R. (2003) PNAS policy on publication of sensitive material in the life sciences. Proc.Natl AcadSci. USA, 100, 1463.

Danchin, A. (2002) Not every truth is good. The dangers of publishing knowledge about potential bioweapons. EMBO Rep., 3, 102-104.

Enserink, M. (2002) Did bioweapons test cause a deadly smallpox outbreak? Science, 296, 2116-2117.

Finkel, E. (2001) Engineered mouse virus spurs bioweapon fears. Science, 291, 585.

Geissler, E. (1999) in Biological and Toxin Weapons:Research,Development and Use from the Middle Ages to 1945 (eds Geissler, E. & Moon, J.E.v.C.), 91-126.
Stockholm International Peace Research Institute, Oxford Univ. Press, Oxford, UK. Geissler, E. & Moon, J.E.v.C. (1999) Biological and Toxin Weapons:Research,Development and Use from the Middle Ages to 1945. Stockholm International Peace Research Institute, Oxford Univ. Press, Oxford, UK.
Harris, S. (1992) Japanese biological warfare research on humans: a case study of microbiology and ethics. Ann.N.Y.Acad.Sci., 666, 21-52.
Harris, S. (1999) in Biological and Toxin Weapons: Research, Development and Use from the Middle Ages to 1945 (eds Geissler, E. & Moon, J.E.v.C.), 12-152.
Stockholm International Peace Research Institute, Oxford Univ.
Press, Oxford, UK.
Harris, S.H. (2002) Factories of Death.Japanese Biological Warfare,1932-1945,and the American Cover-up, revised edn.
Malakoff, D. (2003) Researchers urged to self-censor sensitive data. Science 299, 321. McNeill, W.H. (1976) Plagues and People. Doubleday, New York, USA.
Mervis, J. & Stokstad, E. (2002) Bioterrorism. NAS censors report on agriculture threats. Science, 297, 1973-1975.
Meselson, M. et al. (1994) The Sverdlovsk anthrax outbreak of 1979. Science, 266, 1202-1208.
Miller, J., Engelsberg, S. & Broad, W. (2002) Germs.Biological Weapons and Americas Secret War.Simon & Schuster, New York, USA. Moon, J.E.v.C. (1992) Biological war allegations: the Korean War case. Ann.N.Y.Acad.Sci., 666, 53-83.

BOOKS BY THIS AUTHOR

The Hydrogen Economy: The Fuel Of The Future

Interest in hydrogen is growing, with demand increasing rapidly. It is clear that the next significant transformation in the energy transition will be based on the hydrogen economy, have a brief description of the topic with each and every aspect of hydrogen economy through this book.

Internet Of Things-Iot : Definition, Characteristics, Architecture, Enabling Technologies, Application & Future Challenges

Here's everything in the book you need to know about the increasingly connected world

The future of IoT has the potential to be limitless. Advances to the industrial internet will be accelerated through increased network agility, integrated artificial intelligence (AI), and the capacity to deploy, automate, orchestrate and secure diverse use cases at hyper-scale. The potential is not just to enable billions of devices simultaneously but also to leverage the huge volumes of actionable data that can automate diverse business processes. As networks and IoT platforms evolve to overcome these challenges, through increased capacity and AI, service providers will edge furthermore into IT and web-scale markets – opening entire new streams of revenue.

Solid State Battery: A Battery From Future

Are solid-state batteries the next game-changer?

Get most of the information about the solid-state batteries from their working to their application. The world needs more power, preferably in a form that's clean and renewable.